# Safety at Street Works and Road Works

A Code of Practice issued by the Secretary of State
for Transport, Local Government and the Regions,
The Scottish Executive and the National Assembly for Wales
under sections 65 and 124 of the New Roads and Street Works
Act 1991, and by the Department for Regional Development
(Northern Ireland) under article 25 of the Street Works
(Northern Ireland) Order 1995

Published for the Department for Transport under licence from the Controller of Her Majesty's Stationery Office.

© Crown Copyright 2001.

Third Impression 2004.

All rights reserved.

Copyright in the typographical arrangement and design is vested in the Crown.

Applications to reproduce the material in this publication should be addressed to Her Majesty's Stationery Office, St Clements House, 2-16 Colegate, Norwich NR3 1BQ. Fax: 01603 723000 or email copyright@hmso.gov.uk

First Published 2001

First Edition Crown Copyright 1992

ISBN 0 11 551958 0

Printed in Great Britain on material containing 75% post-consumer waste and 25% ECF pulp.

The Stationery Office 164856 C500 03/04

## Acknowledgements

The Department for Transport, Local Government and the Regions is grateful to the following for their help in preparing this Code of Practice:

Department for Regional Development (Northern Ireland)
National Assembly for Wales
The Scottish Executive
Highway Authorities and Utilities Committee
Local Government Association
National Joint Utilities Group
Association of Chief Police Officers
Health & Safety Executive
and many other organisations and individuals who provided comments and contributions.

# Foreword

Today's roads are full of fast, heavy traffic. Drivers have to keep a constant look-out for changing road conditions. Whilst this code is primarily directed at you the operative, supervisors and managers have an important responsibility to make sure that all street and road works and operatives are safe. Road users should not be put at risk, and should be informed well in advance about the size and nature of any obstruction. This applies to vulnerable users – including pedestrians, cyclists, motorcyclists and horse riders – as well as drivers.

You must also pay particular attention to the needs of blind and disabled people, children, elderly people and people with prams.

This Code of Practice will help you to safely carry out signing, lighting and guarding of street works and road works on most roads (see page 1).

With effect from 1st February 2002 this code has statutory backing for street works in England, Wales and Northern Ireland, and for road works in Scotland, as a Code of Practice under the New Roads and Street Works Act 1991 and the Street Works (Northern Ireland) Order 1995. Failure to comply may lead to criminal prosecution in addition to any civil proceedings.

Department for Transport, Local Government and the Regions
The Scottish Executive
The National Assembly for Wales
Department for Regional Development (Northern Ireland)

## APPLICATION

This Code of Practice is issued by the Secretary of State for Transport, Local Government and the Regions, the Scottish Executive and the National Assembly for Wales under sections 65 and 124 of the New Roads and Street Works Act 1991, and by the Department for Regional Development (Northern Ireland) under article 25 of the Street Works (Northern Ireland) Order 1995. The legislation requires an undertaker, and those working on its behalf, carrying out work under the Act or the Order to do so in a safe manner as regards the signing, lighting and guarding of works. Failure to comply with this requirement is a criminal offence. Compliance with the Code will be taken as compliance with the legal requirements to which it relates.

Highway authorities in England and Wales and roads authorities in Scotland should comply with this Code for their own works, as recommended by the respective national administrations. The Northern Ireland road authority is legally required to comply with the Code. In the application of this Code to Scotland, all references in the text to 'highway authorities' are to be read as references to 'roads authorities'.

Everyone on site has a personal responsibility to behave safely, to the best of their ability. Under the Health and Safety at Work etc Act 1974, employers have duties to protect their employees from dangers to their health and safety, and to protect others who might be affected by the work activity (e.g. passing pedestrians and motorists). These include proper arrangements for design (including planning and risk assessment) and management (including supervision) of the works. Supervisors qualified under the New Roads and Street Works Act or the Order will know what to do in most situations about which they have to be consulted, and will be able to find out quickly what to do about the others. It is the employer's responsibility to ensure that these arrangements are properly carried out.

This Code applies to all highways and roads except motorways and dual carriageways with hard shoulders. More detailed advice, and advice on some situations not covered by this Code, can be found in Chapter 8 of the Traffic Signs Manual published by the Department for Transport, Local Government and the Regions in conjunction with the Scottish, Welsh and Northern Ireland administrations. This gives authoritative advice, but it does not have the status of a Code of Practice under the Act. In Northern Ireland the use of Chapter 8 is mandatory for undertakers' works on motorways or dual carriageways with hard shoulders, and elsewhere in the United Kingdom undertakers should comply with Chapter 8 when carrying out such works. On all other roads they meet their obligations under section 65 or 124 of the Act, or under article 25 of the Northern Ireland Order, if they comply with this Code, even though further relevant advice may be available in Chapter 8 and other relevant documents.

## THE ILLUSTRATIONS

The illustrations show typical layouts, equipment and methods of working. They are not necessarily the only ones which are acceptable. For example, where a vehicle is used on site it may sometimes need to be placed on the other side of the working area, or facing the other direction from that illustrated. Check the text to see what is mandatory and what is optional.

## EQUIPMENT

Traffic signs (including cones, cylinders and red and white barrier planks) and other apparatus used for the control of traffic must conform to the Traffic Signs Regulations and General Directions, or in Northern Ireland, the Traffic Signs Regulations (NI) in force at the time. In respect of other equipment, compliance is achieved by conformity with appropriate European or British Standards where they exist or, alternatively, to a standard or code of practice of a national standards body or equivalent body of any Member State of the European Economic Area, to a relevant international standard recognised for use in any Member State, or to a specification recognised by a competent public authority of any Member State, provided that in-use equivalent levels of safety, suitability and fitness for purpose are met. Equipment of an innovative or traditional nature that does not conform to a recognised standard or specification but which fulfils the purpose provided by the appropriate standards is equally acceptable if, in use, it is safe, suitable and fit for purpose. In judging the suitability of any equipment offered as equivalent, account must be taken of the need on safety grounds to present consistent visual information to users of the highway. Equipment other than traffic signs does not have to match that shown in the illustrations, provided these conditions are met.

# Key Question

Ask yourself this question :

"Will someone coming along the road or footway from any direction understand exactly what is happening and what is expected of them?"

# Contents

Safe works - basic principles. . . . . . . . . . . . . . . 1
Site layout - works area - working space -
safety zone . . . . . . . . . . . . . . . . . . . . . . . . . . . . . 5
What you will need - High visibility clothing -
signs - cones - barriers - information board . . 15
Setting out signs . . . . . . . . . . . . . . . . . . . . . . . 22
Sequence for setting out signs. . . . . . . . . . . . 23
Sign lighting and reflectorisation . . . . . . . . . 26
Works on footways - safe routes for
pedestrians - scaffolding. . . . . . . . . . . . . . . . 28
Footway ramps, footway boards and
road plates. . . . . . . . . . . . . . . . . . . . . . . . . . . 32
Works on two-lane single carriageways . . . . .34
Works on dual carriageways . . . . . . . . . . . . . 38
Works at pedestrian and cycle crossings. . . . 42
Works at road junctions. . . . . . . . . . . . . . . . . 44
Works at permanent traffic signals . . . . . . . . 47
Works at roundabouts . . . . . . . . . . . . . . . . . 49
Cycle lanes and cycle tracks. . . . . . . . . . . . . 51

Traffic Control - need for control - road widths -
setting up - choice of method. . . . . . . . . . . . 52
- Give and Take . . . . . . . . . . . . . . . . . . . 54
- Priority signs . . . . . . . . . . . . . . . . . . . . 56
- Stop/Go Boards . . . . . . . . . . . . . . . . . 58
- Portable Traffic Signals . . . . . . . . . . . . 60
- Stop Works sign. . . . . . . . . . . . . . . . . . 62
Speed control. . . . . . . . . . . . . . . . . . . . . . . . 63
Mobile works and minor works - mobile lane
closures . . . . . . . . . . . . . . . . . . . . . . . . . . . . 66
Works near tramways. . . . . . . . . . . . . . . . . . 68
Works at or near railway level crossings . . . . 70
Reminder. . . . . . . . . . . . . . . . . . . . . . . . . . . 71
Index . . . . . . . . . . . . . . . . . . . . . . . . . . . . . . 72
Distance table . . . . . . . . . . . . .Inside back cover

This Code of Practice anticipates some regulatory changes which will not apply until new Traffic Signs Regulations and General Directions are published in 2002. The existing provisions from the 1994 Regulations, set out below, continue to apply in the meantime:

Information about sign ownership (p12) –
  Traffic cones – not permitted
  Other signs (including barrier boards) – on the back, in characters not exceeding 15mm in height.

Flash rates of road danger lamps (p17) –
  Between 40 and 150 flashes per minute.

Sign lighting (p26) –
  Signs must be lit if speed limit is 40mph or more.

The following signs will be introduced in the new Regulations and are not yet prescribed for general use. STOP – WORKS (pages 26, 27 and 62), CYCLISTS DISMOUNT AND USE FOOTWAY (p51) and convoy working (p63).

# SAFE WORKS - BASIC PRINCIPLES

## Who does what
It is your responsibility to sign, guard, light and maintain your works safely. Take time to plan how you will do this and to decide on what equipment you will need. There will be some pre-planned works where procedures will already have been decided for you.

## Using this code
This Code shows the principles to follow when signing, guarding and lighting works on all highways and roads except motorways and dual carriageways with hard shoulders. It is impossible to illustrate every situation but some of the common ones are shown. Passages in blue ink indicate matters which are the responsibility of supervisors and managers.

Further advice about traffic safety measures for road works is given in Chapter 8 of the Traffic Signs Manual, including for dual carriageways with hard shoulders and motorways. Always consult your supervisor if you are in any doubt about correct procedures or if you are concerned about safety. It is management's responsibility to provide the signs and guarding equipment. It is your responsibility to use them in the right way.

Your supervisor needs to be aware if work is restricted to certain times of the day and whether other conditions may apply.

### On-site risk assessment
To comply with Health and Safety legislation you must carry out an on-site risk assessment to ensure that a safe system of working in respect of signing, lighting and guarding is in place at all times.

### Be seen
Whether on site or visiting, all personnel must wear a high visibility jacket or waistcoat, as appropriate (see page 15). You may also need other protective clothing or equipment for your personal safety.

### Fix signs properly
Signs, lights and guarding equipment must be secured against being blown over or out of position by the wind or by passing traffic. Use sacks at low level containing fine granular material. Alternatively, use equipment having ballasting as part of its construction. Do not use barrels, kerbstones or similar objects for this purpose - they could be dangerous if hit by traffic. Do not use road pins under any circumstances.

Place the first sign far enough from the works to give adequate warning of the hazard (see table inside back cover). Where signs have to be placed on a footway, they should be positioned so as to minimise inconvenience or hazard to pedestrians.

Check regularly that signs have not been moved or damaged or become dirty, including when the site is left unattended for a period of time. Consult your supervisor if the works will make it impossible for drivers to follow a permanent traffic sign. If it needs to be covered, your supervisor will need to notify the highway authority.

### Don't forget the visibility of signs
Signs must be reflectorised unless otherwise indicated (see page 27). Consult your supervisor at times of poor visibility or bad weather conditions as you may need to provide additional signs or to suspend the work. Keep signs clean.

### Traffic on two-way roads
On a two-way road, signs should be set out for traffic approaching from both directions.

### Traffic at junctions
Signs should be set out for traffic approaching from all directions.

### Site layout
You must include the works area, working space and safety zone in the area to be marked off with cones, and lamps placed where necessary. Never use a safety zone as a work area or for storing plant or materials. See inside back cover for the minimum dimensions for the safety zone.

### Additional areas to be signed
If there are any temporary footways in the carriageway, or obstructions such as spoil or plant, which are not already within the working space, sign and guard them separately to the same standard.

### Additional requirements

Sometimes you may have to duplicate the warning signs on both sides of the road. An example of this would be where signs on the left hand side become obscured by heavy traffic. On dual carriageway roads, the warning signs need to be duplicated on the central reservation.

The road width and volume of traffic at the works site may make traffic control necessary. See page 52 for details of which type of control is appropriate.

Drivers visiting the works must switch on their roof-mounted amber beacons, if they have them, before signalling to enter the works. This will help to make sure that other drivers will not be misled into entering the coned-off area as well. Hazard warning lights confuse other road users so don't use them when entering or leaving a site.

### Maintenance of site

Always keep the site safe with signs, cones, lights and barriers clean and correctly placed. When no one is on site, make sure that the site is regularly inspected. Damaged or displaced equipment must be replaced promptly. Emergencies should be dealt with without delay.

### Changing traffic conditions

Where site or traffic conditions change, appropriate adjustments should be made to signing, lighting and guarding.

### Clearing up

On completion of the works, ensure that all plant, equipment and surplus materials are removed promptly from the site. All signs, lighting or guarding equipment should be removed immediately.

## SITE LAYOUT (See diagrams on pages 8 to 14)

### What is the works area ?
The works area is the excavation, chamber opening, etc, at which you will be working.

### What is the working space ?
The working space is the space around the works area where you will need to store tools, excavated material, equipment and plant, etc. It is also the space that you need to move around in to do the job.

You must leave enough working space to make sure that the movement and operation of the plant (e.g. swinging of jibs and excavator arms) is clear of passing traffic and is not encroaching into the safety zone, or adjacent footway or cycle track.

### What is the Safety Zone ?
The safety zone is the zone provided to protect you from traffic and to protect the traffic from you. **You must not enter the safety zone in the normal course of work.** Materials and equipment must not be placed in the zone. **You will need to enter the zone only to maintain cones and other road signs.**

The Safety Zone is made up of :
- **The length of the lead-in taper of cones (T)**
  This will vary with the speed limit and the width of the works.

- **The longways clearance (L)**
  This is the length between the end of the lead-in taper of cones (T) and the working space.
  It will vary with the speed limit.

- **The sideways clearance (S)**
  This is the width between the working space and moving traffic. The sideways clearance is measured from the outside edge of the working space to the bottom of the conical sections of the cones on the side nearest to the traffic (see page 8). It will vary with the speed limit.

- **The exit taper**
  This is always at $45°$ to the kerbline or road edge.

Turn to the inside back cover for dimensions T, L and S.

You must provide working space and safety zones when personnel are present, but when no personnel are on site the width of the zone can be reduced to make it less of an obstruction to traffic. Dimensions L and S can be reduced (or these spaces omitted altogether) and T adjusted to match the reduced width. L, S and T should be restored to the appropriate dimensions when work on site restarts.

Always aim to provide full safety zone clearances consistent with the speed limit in force. To help achieve this, the unobstructed width of road available for traffic may be reduced to the desirable minimum or absolute minimum (see page 52) for the type of situation, but remember to leave enough room for the swept path of large vehicles at junctions and bends, bearing in mind that at widths of 3 metres or less, the wing mirrors of commercial vehicles could easily overhang the footway.

If pedestrians are diverted into the carriageway, you must provide a safety zone at all times between the outer pedestrian barrier and the traffic.

The recommended lead-in taper is given in the table on the inside back cover. This should be used wherever possible. (For example, page 46 shows how this can be done across a junction.) Sometimes it may be impracticable to provide the full taper. If this happens on congested roads with speed limits of 30mph or less, it is permissible to reduce the lead-in taper to an angle of not more than 45° to the kerb, particularly if the parking of vehicles is usual.

The existing speed limit or temporary speed limit approved by the highway authority should be used to determine the appropriate clearances. If traffic consistently exceeds the speed limit, this should be taken into consideration when reviewing the width of the safety zone. If you feel at risk from vehicles exceeding the speed limit, your supervisor should be requested to contact the police.

Where the carriageway width is so restricted as to prohibit the provision of the appropriate sideways clearance detailed above and diversion of traffic would be impracticable, traffic speeds must be reduced to less than 10 mph and an agreed safe method of working imposed on the site. (See page 63)

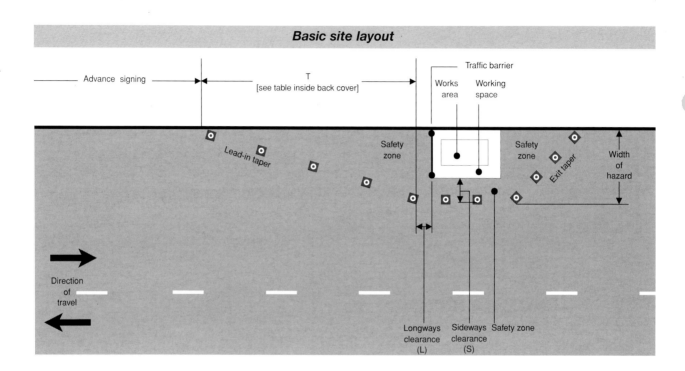

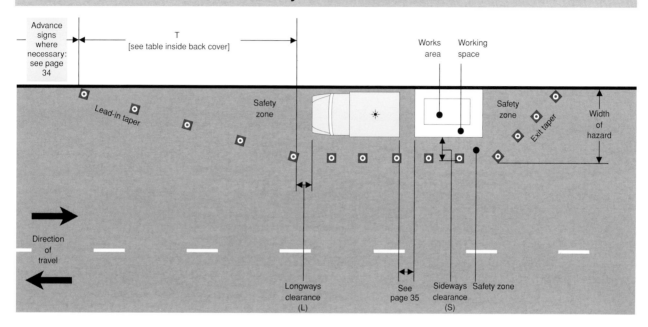

## Basic site layout

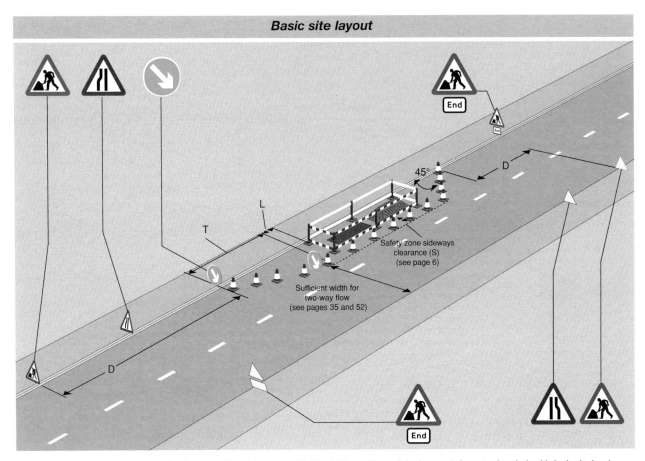

For numbers and size of cones, length of lead in taper (T), and dimensions 'D', 'L' and 'S' see table inside back cover. Information boards should also be displayed (although omitted here for clarity). See page 20.

## Basic site layout with works vehicle

For numbers and size of cones, length of lead in taper (T), and dimensions 'D', 'L' and 'S' see table inside back cover. Information boards should also be displayed (although omitted here for clarity). See page 20.

11

## Basic signs and equipment you will need

Road works ahead

Road narrows on left-hand side ahead

Road narrows on right-hand side ahead

Keep right

Keep left

Traffic cone

Road danger lamp

Pedestrian barrier (see page 18)

Traffic barrier ('Lane closed' sign)

End of road works

You will also need an information board

High visibility clothing (see page 15)

Information about the ownership of traffic signs may be shown as follows:
Traffic cones: embossed on the base in the same colour as the base, in characters not more than 80mm high.
Barrier planks: indicated on the back in characters not exceeding 50 millimetres in height where they are in a contrasting colour, or 80mm in height where they are embossed in the same colour.
Other signs: indicated on the back in characters not exceeding 25mm in height, where they are in a contrasting colour, or 50mm in height where they are embossed in the same colour.

## Some other signs you may need for which you should refer to your supervisor

Additional signs may be needed according to the circumstances. Some of the more common situations are dealt with in greater detail in the following pages.

 Road narrows on both sides ahead

 Traffic signals ahead

 Where vehicles should stop at temporary traffic signals

 Traffic control ahead

 Stop/Go boards

 Priority to vehicles from opposite direction

 Priority over vehicles from opposite direction

 Other danger ahead (use only with a plate)

 Slippery road

 Traffic cylinder

 High intensity flashing beacon

 Sharp deviation to the right

 Direction of temporary pedestrian route

 Stop Works

 Traffic signals not in use

Or variations of these signs

 Temporary road surface

 Zebra or signal controlled crossing is not in use

 Loose chippings

 Ramp ahead

 Ramp

 Left-hand lane of a dual two-lane carriageway road closed

 Left-hand lane of a dual three-lane carriageway road closed

 Centre lane of a three-lane two-way road closed

 Cyclists dismount

 Setting out road works ahead

 Single file traffic

 Overhead cable repairs / Surveying / Overhead works

 Distance over which hazard or prohibition extends

 Distance to hazard or obstruction

 Distance and direction to hazard or obstruction

 Maximum speed advised

The signs shown on page 12, 13 and 14 may be bilingual where permitted.

## Some permitted combinations of the signs and plates

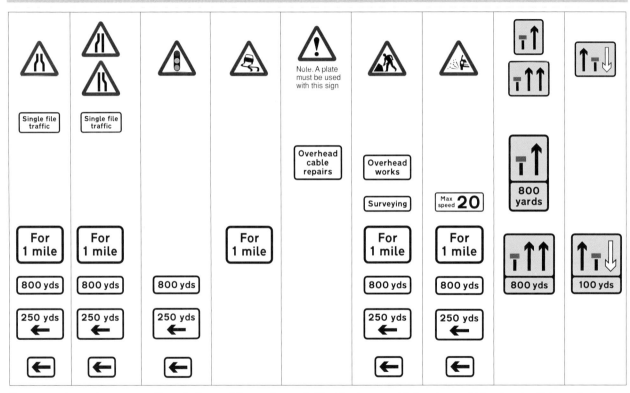

Distances may be varied as appropriate. Except in the case of distance plates used with wicket signs on mobile lane closure vehicles, plates should always take the background colour of the main sign.

## WHAT YOU WILL NEED

It doesn't matter whether the works are small or large, on the ground or overhead, all street works require warning and information: for the basic layout see page 5. In emergencies as much warning must be given as the circumstances permit, and full signing must be provided as quickly as possible.

### *High visibility clothing*

You will have been provided with High Visibility Clothing conforming to BS EN471:1994, Table 1, Class 2 or 3, which must be worn at all times. It will comply with the requirements of Clause 4.2.3(b) in all cases.

Jackets with sleeves in accordance with Clause 4.2.4 and to Class 3 must be worn on dual carriageway roads with a speed limit of 50 mph or above, unless operatives stay within the working space at all times. The colour of the background material should normally be fluorescent yellow from table 2 of BS EN471:1994, and the retroreflective material should comply with Table 5.
High Visibility Clothing to BS6629:1985 may continue in use until 30th September 2002.

### *Advance signs*

These should be placed where they will be clearly seen, and cause minimum inconvenience to drivers, cyclists, pedestrians and other road users alike, and where there is a minimum risk of their being hit or knocked over by traffic. Where there is a grass verge the signs should normally be placed there; the placing of signs in the footway is permitted but in no circumstances must the footway width be reduced below 1 metre.

If there are already vehicles parked in the carriageway, place the advance signs so that they are not obscured.

The **'Road Works Ahead'** sign is the first sign to be seen by the driver, so place it well before the works. Its size, the minimum distance from the start of the lead-in taper, and clear visibility distance will vary according to the type of road and its speed limit - see table on inside of back cover. The range of distances is given to allow the sign to be placed in the most convenient position bearing in mind available space and visibility for drivers. Do not simply choose the minimum distance - assess each site carefully.

A **'Road Narrows Ahead'** sign warns the driver which side of the carriageway is obstructed. Place it midway between the Road Works Ahead sign and the beginning of the lead-in taper.

On roads with speed limits of 50 mph or more, all advance signs should have plates giving the distance to the works in yards or miles (not in metres).

### Keep Right and Keep Left signs

Place **'Keep Right'**, or as appropriate, **'Keep Left'** signs at the beginning and end of the lead-in taper of cones. These signs must be the same size as the Road Works Ahead sign. MAKE SURE THAT THE SIGNS POINT IN THE CORRECT DIRECTION. Do not turn the sign frame on its side to make it point in the correct direction. These signs must NOT be used for directing pedestrians.

## Cones and lamps

Place a line of Traffic Cones to guide traffic past the works and add Road Danger Lamps in poor daytime visibility and bad weather. Where the traffic is faster the length of taper must be longer. Look at the table inside the back cover for details of positioning of cones and lamps.

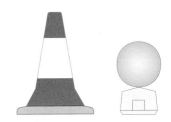

Road Danger Lamps must be used at night on roads with a speed limit of 40 mph or above. On roads with a lower speed limit, judgement may be used as to whether Road Danger Lamps are needed, depending on the standard of street lighting.

Road Danger Lamps must not be higher than 1.5 metres above the road (or 1.2 metres where the speed limit is more than 40 mph).

The type of lamp to be used is as follows :

| Type of Road Danger Lamp | Conditions of use |
|---|---|
| Flashing lamp (55 to 150 flashes per minute) | Only when ALL of the following conditions apply :<br>- the speed limit is 40 mph or less<br>- the Road Danger Lamp is within 50 metres of a street lamp, and<br>- the street lamp is illuminated. |
| Steady lamp | On any road with or without street lighting. |

## Barriers

Barriers may comprise separate portable post and plank systems, 'gate frames' linked together, or semi-permanent constructions built to enclose the site.

There are several different requirements for the barrier planks associated with post and plank systems. The following explains the requirements and how they may be met using barrier planks which are red and white and manufactured in fully retroreflective materials. (Note: 'Retroreflective' means that at night the material reflects light back to the light source).

Barrier planks are required to carry out three functions :-

1. As a **TRAFFIC BARRIER**. When a traffic lane is closed for works to take place, the regulations require this to be done with a retroreflective red and white barrier plank placed across the lane. This is illustrated on page 12 as a Traffic Barrier ('Lane Closed' sign).

2. As a **PEDESTRIAN BARRIER**. Pedestrians must be separated from the works by barriers which are conspicuous and mounted as part of a portable fencing system. Pedestrian barrier planks may be of several different contrasting colours; yellow, white or orange colours are best detected by partially sighted people, but red and white is one of the acceptable combinations.

3. As a **TAPPING RAIL** for blind and partially sighted people. Tapping rails are placed as the bottom rail in a pedestrian fencing system. A red and white barrier plank may be used.

All barriers facing vehicular traffic should be of the fully retroreflective red and white form. Red and white barrier planks do not have to be used for pedestrian barriers or tapping rails but, if they are, they must be retroreflective. Other planks used for these purposes do not need to be retroreflective.

There are other points to note about the use of barrier planks in portable fencing systems:
   a) The TRAFFIC BARRIER ('Lane Closed' sign) is not needed if the works are protected by a conspicuous vehicle.
   b) Pedestrian barrier systems must be rigid enough to guard pedestrians from traffic, excavations, plant or materials. They must be placed with sufficient clearance to prevent pedestrians falling into the excavation and, when placed to create a temporary footway in the carriageway parallel to the traffic stream, must be protected by a row of traffic cones between the barrier and the traffic stream. Consult your supervisor if the excavation is deep, or positioned close to pedestrians, as stronger barriers may be needed and/or other safety measures may be required e.g. covering or temporarily refilling the excavation.
   c) Where a work site may be approached by pedestrians crossing the road from the opposite side, you should place barriers, including tapping rails, all around the excavation, even when pedestrians are not diverted into the carriageway.
   d) Where long excavations are sited in situations where pedestrians are not expected to cross from the opposite side, barriers on the traffic stream side of the works area do not need the tapping rail. In these circumstances, on an unrestricted road, the barrier on the traffic stream side can be replaced with an additional row of cones. These cones should be linked with a suitably supported traffic tape to attract attention to the boundary of the safety zone.

Use pedestrian barriers to mark out any temporary footway. You must always use a rigid barrier to protect pedestrians from traffic, excavations, plant or materials. Place road danger lamps at the ends of the barriers at night so that they may be clearly seen by pedestrians.

PORTABLE PEDESTRIAN BARRIERS, which may include mesh, should be reasonably rigid, designed to resist being blown over by the wind or passing traffic, and have:
- a handrail fixed at between 1 metre and 1.2 metres above ground level, which should be reasonably smooth and rigid to guide pedestrians and give them some measure of support; and a visibility panel of at least 150mm deep which may be integral with the handrail or, if separate, must be fixed so that its upper edge is a minimum of 0.9 metres above ground level.
- a tapping rail (or equivalent reasonably rigid area if the barrier is a vertically continuous one) of minimum depth of 150mm with a lower edge at ground level or set at up to 200 mm above ground level.

## Information board

An information board must be displayed at every site, except for mobile works and minor works which do not include excavation, involving use of a vehicle (see page 66). This board should be placed so that it does not obstruct footways or carriageways but can be read mainly by pedestrians, and possibly by drivers who have stopped. (The boards are too detailed to be read easily by passing traffic.)

The board must give the name of the organisation for which the works are being carried out, and a telephone number which can be contacted in emergencies. It may also contain other information that will be helpful in explaining to the public why the work is being done, who is doing it and how long it will take. Such additional information is to be encouraged where practical and could include some or all of the following: a brief description of the works, the name of the contractor and a message apologising for inconvenience or delays. A completion date should normally be included if the works are expected to continue for more than a month.

> R<sub>W</sub> Riversford Water Company PLC
> 
> **WATER MAIN RENEWAL**
> Completion expected
> September 2002
> Contractor - N E Samson Ltd
> **Emergency Telephone**
> **020 7123 4567**
> 
> Sorry for any inconvenience

## End sign

The 'End' sign indicates not only the end of works but also the end of any temporary restrictions, including temporary speed limits, associated with the works.

If the permanent speed limit changes within the length of road covered by a temporary speed restriction, signs indicating the new speed limit must be provided on each side of the carriageway at the end of the works, in addition to the 'End' sign.

You must place an 'End' sign beyond works that are 50 metres or more in length (measured between the end of the lead-in taper and the beginning of the exit taper) and beyond two or more adjacent sites.

But an 'End' sign is not necessary on a road where ALL of the following conditions are met:

- there are no temporary speed limits or other traffic restrictions
- the speed limit is 30 mph or under
- there is a total two-way traffic flow of less than 20 vehicles counted over 3 minutes (400 veh/hr)
- less than 20 heavy goods vehicles pass the works site per hour.

# SETTING OUT SIGNS

## *Parking*
You must park your vehicle safely before you unload or set up signs. If you can't park it off the road make sure the vehicle can be seen clearly by other drivers. Turn on your roof-mounted amber beacon(s). Do not obstruct a footway or cycle track when parking off the road.

## *Precautionary measures*
If at all possible place signs so that they do not obstruct vehicles, cyclists, pedestrians or other road users. Where there is no street lighting, place a Road Danger Lamp alongside signs which are on, or partly on a footway to warn pedestrians at night.

## *Advance signing*
It is important that the distances, including the safety zone dimensions, are determined before starting to set the signs out. From the table inside the back cover select the distance for the advance signs. If there is limited visibility on the approach to the proposed works site, e.g. on a bend, on a dip in the road or on the brow of a hill, you must provide extra advance signs. These extra signs will need to be placed first.

## *Access to works site*
You may need to allow for vehicles entering and leaving the site.

# SEQUENCE FOR SETTING OUT SIGNS

## Setting out advance signing where necessary

When the Road Works Ahead sign is more than $1/4$ mile from the works, or when extra advance signs are needed because visibility is limited :

1. Stop the vehicle in a safe place, switching on your roof-mounted amber beacon(s)
2. Ensure you are wearing your high visibility clothing
3. Set up the signs you need before moving on to the works site and setting out the rest of the layout.

## Setting out basic signing

You are at greatest risk when setting out the signing and guarding, so great care is needed to ensure that you can see the traffic and the traffic can see you.

Wear your high visibility clothing, putting it on before leaving the vehicle. It may be safer to get out of the vehicle on the passenger side, rather than stepping into the traffic stream.

Make sure the roof-mounted amber beacon(s) are switched on and operating.

Face the traffic when setting out signs, taking particular care when you are crossing the road to place signs.

Follow the sequence on pages 24 and 25.

If you can, you must park your vehicle in a safe place. If you park in the road, you must protect it from traffic going past.

Set up a 'Keep Right' sign at the outside corner of the vehicle, along with a Traffic Cone.

Set out the 'Road Works Ahead' sign at the distance which you have already decided on ('D' from the table inside the back cover).

Measure or pace out the distance. Then put one sign on the left-hand side, and another on the other side of the road if required.

Using the diagrams to help you, work back towards the site placing more signs as necessary. Keep on the verge or footway if you can.

If you are on a two-way road repeat this procedure and place signs for traffic going in the opposite direction. If portable traffic signals or stop/go boards are needed, start using them now. Then cone off the works area.

Always face the traffic when you set out the cones for the lead-in taper. Start from the kerb or road edge. Complete the coning around the works, leaving enough room for working space and safety zones.

Use cones, 'Keep Right' signs, barriers and lamps, and Information Board to complete the warning, guidance and protection for the works.

Where appropriate set up 'End of Road Works' sign to show that the road is clear in both directions.

**When you need to remove the signs, reverse the procedure shown here.**

# SIGN LIGHTING AND REFLECTORISATION

## Stop/Go Boards and Stop-Works sign
You must always directly light Stop/Go Boards and Stop-Works sign at night.

## Sign lighting

All variations of these signs

These signs, and any plates used with them, must be directly lit when ALL of the following conditions apply :
- permanent speed limit of 50 mph or above
- there is general street lighting
- the street lighting is on
- the sign is within 50 metres of a street light

### Reflectorisation

All signs (including traffic cones, cylinders and red and white barrier planks), except the 'Direction of Temporary Pedestrian Route' sign, 'Crossing not in use' sign, the 'Pedestrians look left' sign and the Information Board, must be reflectorised to BS873 : Part 6 : 1983 Class 1 or Class 2 or BS873 : Part 8 : 1985 designation 1 or 2, or the equivalent standard of a European Economic Area State. It is unlawful to reflectorise only part of the sign face; where signs are required to be reflectorised, this applies to the whole surface except for any part coloured black. Non-reflectorised elements of a sign face appear black at night; this not only reduces the sign's visibility, but may well make its message unrecognisable.

The 'STOP-WORKS' sign must be reflectorised. The part coloured yellow must also be fluorescent and the part coloured red may be fluorescent.

# WORKS ON FOOTWAYS - LOOK AFTER PEDESTRIANS

## Pedestrian safety

It is your responsibility to make sure that pedestrians are safe during the works. This means protecting them from both the works and passing traffic.

You **must** take into account the needs of children, elderly people and people with disabilities, having particular regard for visually impaired people. In order to do this you must provide a suitable barrier system (see page 18) which safely separates pedestrians from hazards and provides sufficient access for people using wheelchairs and those with prams or pushchairs.

## Safe routes for pedestrians

If you have to close a footway or part of a footway, you must provide a safe route for pedestrians which should include access to adjacent buildings, properties and public areas.

Safe routes should always provide a minimum unobstructed width of 1 metre, increased wherever possible to 1.5 metres or more.

However, a balanced assessment must be made to provide pedestrians with the safest option. For example, a route of 1 metre unobstructed width which uses the existing footway is potentially safer than a wider temporary route placed in the carriageway.

When temporary pedestrian routes have to be placed in the carriageway, make sure the signing and guarding barriers are put into place before the footway is blocked. Make sure the sideways clearance (S) of the safety zone is on the traffic side of the barriers. Where necessary, provide ramps, or a raised footway or boards which are fit for the purpose (see page 32).

The use of the other footway may be acceptable in some quiet roads, but if you select this option you must ensure that the alternative route is safe to use, and you must take account of the needs of children and people with disabilities.

## *Pedestrian crossings and pedestrianised areas*

If the works are on or near pedestrian crossings turn to page 42 for advice. In pedestrianised areas the working space, including vehicles, plant or materials, must be enclosed with pedestrian barriers.

## *Safety zone for operatives*

When working in a footway remember you must provide a safety zone in the carriageway if the working space is closer to the edge of the carriageway than the width of the sideways clearance (S). If cones are placed in the road, signing and guarding of the safety zone must be carried out.
These same principles apply when working in a verge adjacent to the carriageway.

## *Scaffolding*

Any scaffolding erected in the highway must be licensed by the highway authority.

## Works on footways

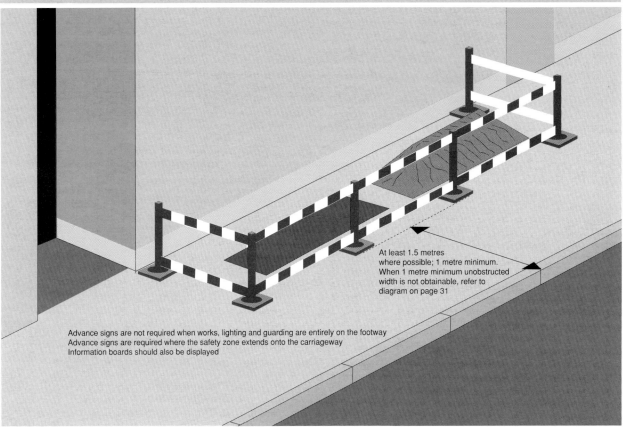

At least 1.5 metres where possible; 1 metre minimum. When 1 metre minimum unobstructed width is not obtainable, refer to diagram on page 31

Advance signs are not required when works, lighting and guarding are entirely on the footway
Advance signs are required where the safety zone extends onto the carriageway
Information boards should also be displayed

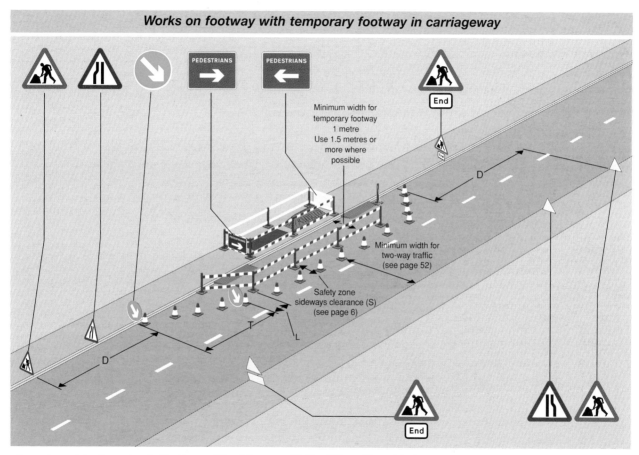

For numbers and size of cones, length of lead in taper (T) and dimensions 'D', 'L' and 'S' see table inside back cover. Information boards should also be displayed (although omitted here for clarity). See page 20.

# FOOTWAY RAMPS, FOOTWAY BOARDS AND ROAD PLATES

## *Footway Ramps*

When pedestrians are diverted to temporary footways in the carriageway, suitable ramps must be provided to enable people using wheelchairs or pushchairs to negotiate kerbs safely.

Ramps should cover the full width of the temporary footway (minimum of 1 metre), and should be constructed from materials strong enough to support pedestrians, preferably with edging to prevent wheelchairs slipping over the edge. They may be made on site, eg. from wood or bitumastic materials, or prefabricated. Ramps should slope gently enough to enable wheelchair users and pushchairs to reach the kerb without undue difficulty. Ideally, the layout should include a platform at kerb level which would allow wheelchair users to turn through 90° before descending the ramp in a line that is parallel to the kerb. Ramps must be fixed in position, allow for rain water to run along the gutter, and should have a slip resistant surface.

## *Footway Boards*

Footway Boards should only be used on footways to maintain foot and light vehicle access to premises during excavation works.

Footway Boards used for bridging excavations must provide at least 1 metre width for pedestrians, but preferably 1.5 metres, must be strong enough to support pedestrians, and must be made from material which is unlikely to become distorted. Where used for light vehicles the boards must be capable of supporting the added load and, where used on a vehicle crossover the whole width of the crossover must be boarded.

The edges of footway boards must be chamfered to prevent tripping and should have a slip resistant surface. The sides of the excavation should be stable or suitably supported under the board, and the board should be rigidly fixed with sufficient length beyond the excavation to provide the necessary support. The edges of the footway boards adjacent to the excavation should be fenced to prevent falls.

### Road Plates

Road Plates may be required to bridge excavations in order to open the carriageway to traffic, e.g. during traffic sensitive periods, at night or at weekends. Their use must always be authorised by your supervisor who will decide on the size and thickness of the plate to be used. The thickness will depend on the width of excavation to be spanned and the type of traffic expected to use them.

Road Plates must be made of suitable material with an appropriate skid resistant surface. Their installation must not present a hazard to cyclists or motorcyclists.

The sides of the excavations must be suitably supported beneath the road plates, and they must be rigidly secured to the road surface. Road Plates must be either sunk into the surface or suitable bitumastic material used to provide a ramp to the plate level. Where ramps are used, appropriate Ramp Warning signs should be used when there is a significant change in the road level.

The edges of the Road Plates adjacent to the excavation should be fenced to prevent falls.

## WORKS ON TWO-LANE SINGLE CARRIAGEWAY ROADS

Use the basic layout on pages 8-11 for the approach signing and guarding.
You may omit the traffic barrier if the works are protected by a conspicuous vehicle.

Exceptions are allowed in either one of cases (a) and (b) below, provided that ALL the following conditions are met :
- the speed limit is 30 mph or under
- there is a total two-way traffic flow of less than 20 vehicles counted over 3 minutes (400 veh/hr)
- less than 20 heavy goods vehicles and buses pass the works site per hour.

(a) **Where there is a works vehicle with a roof-mounted beacon in continuous use**
So long as drivers can see the beacon clearly from at least 50 metres in either direction, you do not need to use the 'Road Works Ahead' and 'Road Narrows' signs in advance of the works. However, you must still use the lead-in taper of traffic cones and the 'Keep Right/Left' sign. You will also need to use a Traffic Barrier unless your works vehicle is conspicuously coloured.

(b) **On roads where the parking of vehicles is usual and parked vehicles other than works vehicles are likely to be present for the duration of the works**
Works which take place in the space between parked vehicles need no advance warning, provided that the whole works, including the safety zone, do not extend into the carriageway beyond the line of vehicles. There should be a line of cones in place at all times on the road side of the works. Provision must be made for the possibility that the parked vehicles will be moved. Should this happen, it will be necessary to revert to the basic layout.

## Works vehicle

If you want to park a works vehicle in front of the works to give some physical protection, or to work from it, keep a distance between the vehicle and the works of :
      Speed limit of 30 mph or under - 2 metres
      Speed limit of 40 mph or above - 5 metres

The existing or approved temporary speed limit should be used for the above (but see also page 7). Measure the longways clearance (L) from the end of the lead-in taper to the part of the vehicle which faces the traffic. If you are working from the back of the vehicle, park it so that its back is facing the works.

## Road widths

Turn to page 52 for the minimum road width required for two-way working. If there is not enough space for two-way traffic, it may be possible to use traffic control. Your supervisor will decide for you in consultation with the highway authority.

## Setting out

Turn to the inside back cover for dimensions D, T, L and S. Then turn to page 22 for the setting out procedure.

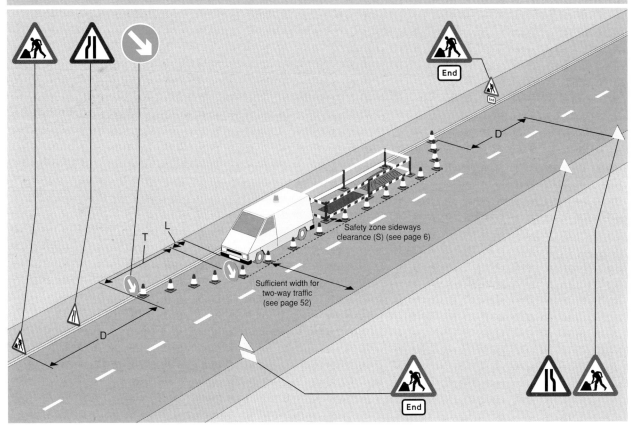

For numbers and size of cones, length of lead in taper (T) and dimensions 'D', 'L' and 'S' see table inside back cover. Information boards should also be displayed (although omitted here for clarity). See page 20.

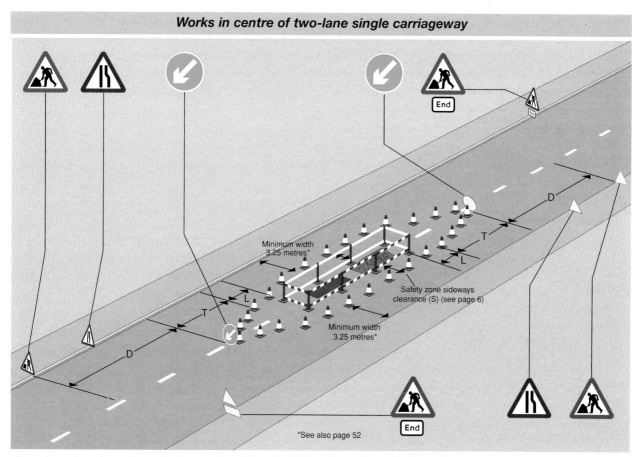

For numbers and size of cones, length of lead in taper (T) and dimensions 'D', 'L' and 'S' see table inside back cover. Information boards should also be displayed (although omitted here for clarity). See page 20.

## WORKS ON DUAL CARRIAGEWAY ROADS WITHOUT HARD SHOULDERS

Consult your supervisor about what to do. Further advice is available in the Traffic Signs Manual Chapter 8. If the work involves closing the right-hand lane, or closing a left hand lane on a three-lane dual carriageway with the national speed limit, you will need to liaise with the highway authority. Speed restrictions may be necessary for which you will also need to refer to the highway authority.

### *Right lane closure*
This is shown on page 40.

### *Left lane closure*
If the left lane is closed, you should normally merge traffic into the left lane by using a guide island, and then divert the traffic into the right lane(s). This is shown on page 41. If two or more lanes are to be diverted to the right, you must change the road markings to make sure that the traffic lanes are continuous.

Where conditions permit you may want to merge traffic to the right at a left lane closure - for example from lane 1 into lane 2. You can do this when :
- lane 1 of a three-lane dual carriageway is being closed, or
- there will be no more than about 60 vehicles per 3 minutes (1200 veh/hr) on each traffic lane which remains open.

## Signing requirements

| Speed limit | Signing requirements |
|---|---|
| 30mph | as shown on pages 40 and 41 except that distance plates may be omitted |
| 40mph | as shown on pages 40 and 41 |
| 50 mph and over | as shown on pages 40 and 41, but with advance warning signs added on each side of the carriageway as shown at the bottom of this page. |

## Setting out

Turn to inside the back cover for dimensions D, T, L and S.
The length of the guide island (G) (see page 41) should be :
- 50 metres for roads with a speed limit of 50 mph or under
- 100 metres for roads with a speed limit of 60 mph or above.

If you think Mobile Lane Closure methods may be required when setting out guarding, you should consult your supervisor.

### Sequence of signing for 50mph or above

*The road works signs at 1 mile from the works are only required on unrestricted roads where congestion is likely.

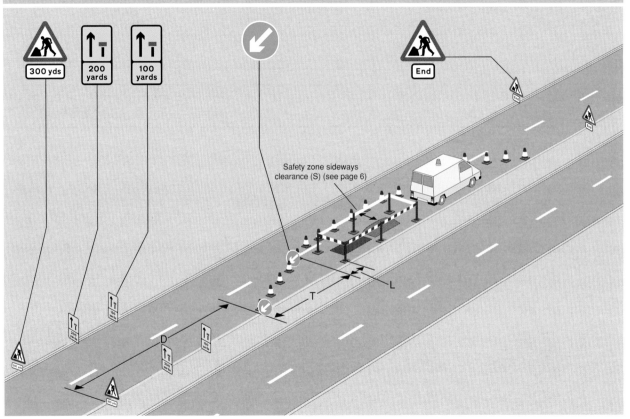

For numbers and size of cones, length of lead in taper (T) and dimensions 'D', 'L' and 'S' see table inside back cover. Information boards should also be displayed (although omitted here for clarity). See page 20.

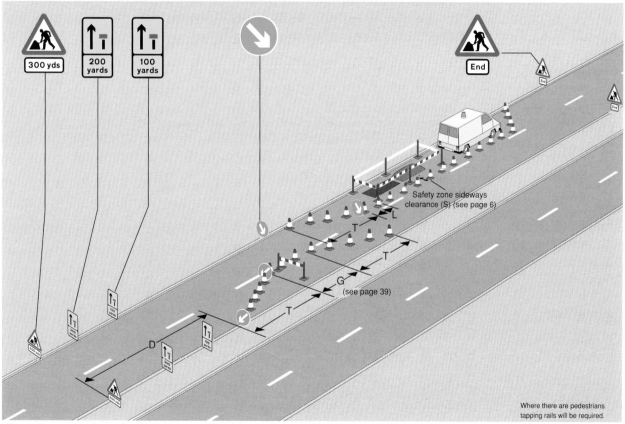

For numbers and size of cones, length of lead in taper (T) and dimensions 'D', 'L' and 'S' see table inside back cover. Information boards should also be displayed (although omitted here for clarity). See page 20.

## WORKS AT PEDESTRIAN AND CYCLE CROSSINGS

Before any work takes place at or near a pedestrian or cycle crossing you must consult your supervisor. Only the highway authority can authorise a crossing to be taken out of service. Where appropriate, alternative signed routes should be agreed with the highway authority.

If due to works the pedestrian or cycle crossing has to be closed, you should :
- ensure the closure has been authorised by the highway authority
- erect 'Crossing not in use' signs
- cover zebra crossing globes or signal heads/buttons, and any other signal operating device
- cover (or arrange with the highway authority to remove) tactile indicators so that visually impaired and deaf people are not misled, especially where tactile paving has been laid
- at signal-controlled crossings, erect 'Traffic signals not in use' signs and cover the signal heads
- if the limits of the crossing are obstructed, or the visibility between drivers and pedestrians/cyclists is reduced to an unacceptable degree, erect barriers across the accesses to the crossing.

Close both crossings if the works spread into one or both sides of a double crossing which has a central refuge.

### *Setting out*
See page 43 for an illustration of a closed signal controlled crossing.

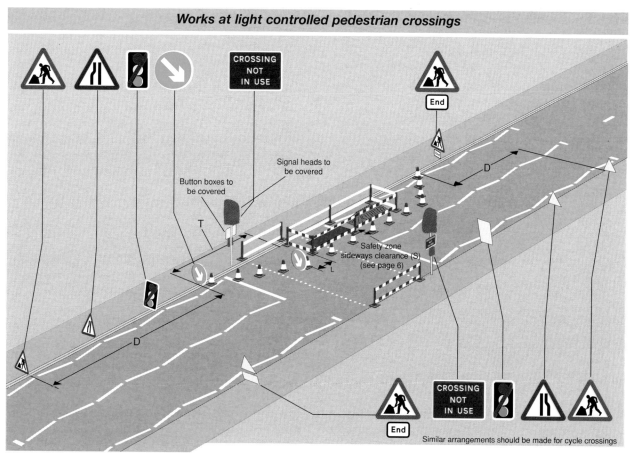

For numbers and size of cones, length of lead in taper (T) and dimensions 'D', 'L' and 'S' see table inside back cover. Information boards should also be displayed (although omitted here for clarity). See page 20.

## WORKS AT ROAD JUNCTIONS

Keep the two-way traffic flowing past the works if possible. If you can't, traffic control or a diversion may be required.

Put up 'Road Works Ahead' signs with arrow plates on the main route if the works are in a side road. Turn to page 45 for details.

The illustration on page 46 shows the works on or near the far side of a junction. At works like these, take the taper of cones up to the approach side of the junction. Make sure that any cones near the junction help drivers turn left smoothly.

## Works at road junctions (1)

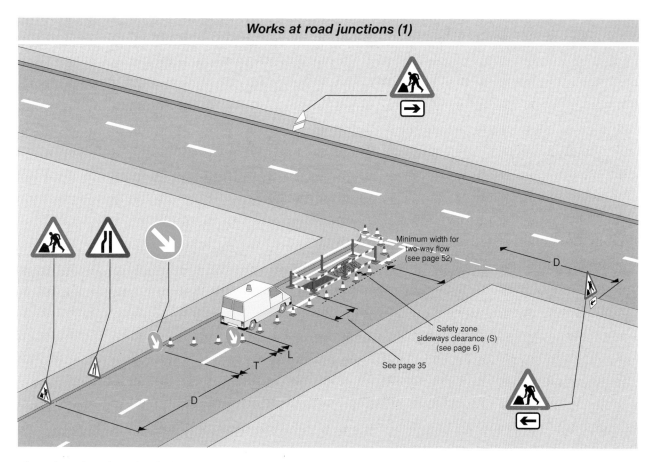

For numbers and size of cones, length of lead in taper (T) and dimensions 'D', 'L' and 'S' see table inside back cover. Information boards should also be displayed (although omitted here for clarity). See page 20.

## Works at road junctions (2)

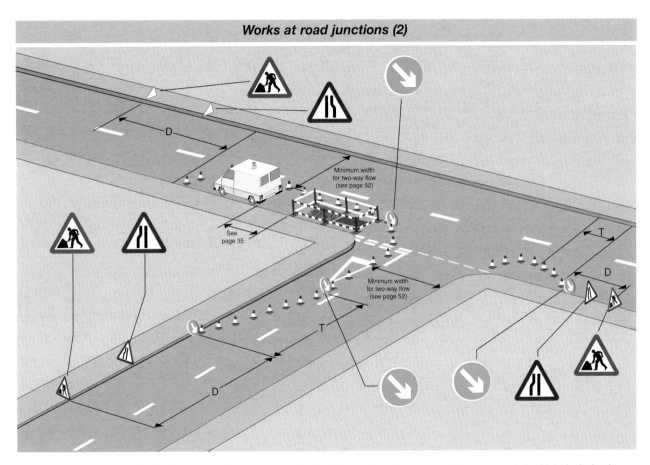

For numbers and size of cones, length of lead in taper (T) and dimensions 'D', 'L' and 'S' see table inside back cover. Information boards should also be displayed (although omitted here for clarity). See page 20.

# WORKS AT ROAD JUNCTIONS CONTROLLED BY PERMANENT TRAFFIC SIGNALS

## Approaches to junctions

A works site on the approach to a traffic signalled junction can cause significant disruption to the traffic flow at the junction. An adjustment of the traffic signals may be required, so consult your supervisor, who will then consult the highway authority.

## At junctions

If traffic signals are not working, put up 'Traffic signals ahead not in use' signs on all approaches. Permanent traffic signals are often replaced by temporary or portable traffic signals for the duration of the works. Both will need approval by the highway authority.

If pedestrian lights at a junction are affected by the works, they should be treated in a similar manner to pedestrian crossings (see page 42). This must be discussed with the highway authority.

Figures 1, 2 and 3 on page 48 show guarding and signing for works where signals are on single carriageway urban roads with a speed limit of 30 mph or under :

- Figure 1 : Work at an island signal when there is no works vehicle present.
- Figure 2 : Work at an island signal when a works vehicle is present, and its operating roof-mounted amber beacon can be seen clearly.
- Figure 3 : Work at a kerbside signal when a works vehicle and its operating roof-mounted amber beacon can be seen clearly.

Appropriate advance warning signs should be placed on the cross arms of the junction.

Consult your supervisor where the road has a speed limit of 40 mph or above.

## Works at traffic signals

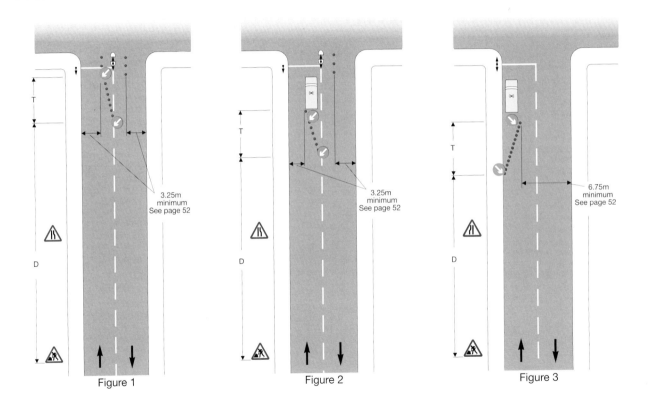

Figure 1  Figure 2  Figure 3

# WORKS AT ROUNDABOUTS

## *Works at the entrance to or exit from a roundabout*
Use advance signs to warn traffic on all approaches that there are works at or near the roundabout. Use 'Keep Right/Left' signs to guide traffic around the coned-off works site.

Try to keep two-way traffic flowing if possible, but remember the width restrictions (see page 52). However, if the works site makes the road too narrow to allow two-way traffic to pass, restrict the road to 'Exit only' from the roundabout. **The traffic usually entering the roundabout on this road will then need to be diverted.** This requires permission of the highway authority and needs to be pre-planned as adequate notice has to be given. Consult your supervisor.

Extra cones will be needed to restrict traffic to one lane going towards this exit and additional advance warning using 'Road Narrows' signs provided on all approaches. Use 'Keep Right/Left' signs to guide traffic past coned areas.

## *Works in the circulatory area of a roundabout*
Movement of traffic should be maintained if possible. Guard and cone the works and provide advance 'Road Narrows' signs on all approaches. Use 'Keep Right/Left' signs to guide the traffic past the works site.

Where works will completely obstruct the circulatory area of a roundabout, consult your supervisor who will then consult the highway authority.

Varying the number of lanes on the circulatory section of a roundabout can distract drivers, therefore consideration should be given to coning down to the same number of lanes unless the traffic pattern dictates otherwise. Lane dedication signs may be needed. Vehicle turning paths need to be carefully considered to ensure the rear wheels of long vehicles do not hit the cones and to ensure adequate width on the restricted approach.

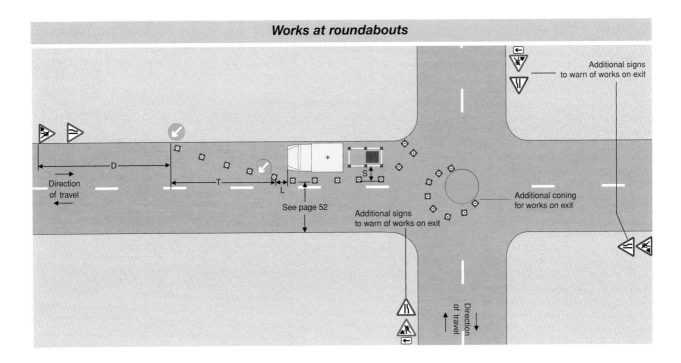

## CYCLE LANES AND CYCLE TRACKS

Where cycle lanes, cycle tracks and cycle routes are affected by street works and road works you should use your best endeavours to ensure the safety of cyclists passing or crossing by the works.

Cycle lanes marked with a solid white line have been created by means of Traffic Regulation Orders. Where one of these is affected by planned works, your supervisor will need to discuss the situation with the highway authority well before the work starts. It may be necessary to obtain a Temporary Notice or Traffic Regulation Order to suspend the cycle lane. Temporary Notices and Temporary Traffic Regulation Orders are not required for emergency works.

Cyclists may have to use the remainder of the carriageway, use an alternative route or, if an alternative route is not available, will have to dismount while passing the works. Your supervisor may need to discuss these alternatives with the highway authority.

When portable traffic signals are used, bear in mind when adjusting the timings that cyclists take longer to clear the controlled section than motor vehicles.

Where the carriageway is closed off but the footway remains open, cyclists should be advised to dismount by using a 'Cyclists Dismount and Use Footway' white on red temporary sign.

Wherever possible, a minimum lane width of 3.25 metres should be provided to allow a car to overtake a cyclist, more where lorries or buses will be present (see page 52).

## TRAFFIC CONTROL

### The need for traffic control

Adequate width is required for two-way working. An unobstructed width of the road past the works must meet the minimum shown in the following table, dependent on the type of traffic. Where possible, widths above the minimum should be provided in order to give space to cyclists (see page 51).

Where these widths cannot be met, **shuttle working** with traffic control must be introduced, and the unobstructed width reduced to a maximum of 3.7 metres.

Where an unobstructed width of at least 6.75 metres for two-way traffic cannot be provided there may be problems for HGVs and buses. In these circumstances, where a bus route exists, you must consult your supervisor who will advise the public transport co-ordinator.

|  | Normal traffic including buses and HGV | Cars and light vehicles only |
|---|---|---|
| Two-way working | 6.75 metres minimum | 5.5 metres minimum |
| Shuttle working with traffic control | 3.7 metres maximum<br>3.25 metres desirable minimum<br>3.0 metres absolute minimum | 3.7 metres maximum<br>2.75 metres desirable minimum<br>2.5 metres absolute minimum |

Where the **absolute minimum** cannot be met, your supervisor must consult the highway authority.

### Setting up traffic control

When setting up traffic control, the cone taper should be at $45°$ to the road edge. Long sections of narrow lanes can cause difficulties for cyclists and horse riders, and this should be taken into consideration where appropriate.

## Choice of traffic control method

Select the method to be used from the table. Check pages 54 to 62 for precise details of the chosen method. If the situation is not covered by the methods shown, your supervisor should consult the highway authority.

| Method | Max Speed limit | Coned area length | Traffic flow (maximum) | Notes |
|---|---|---|---|---|
| Give and take | 30 | 50 metres maximum | 20 vehicles over 3 mins and 20 HGV per hour | Signing as per page 54-55 |
| Priority | 60 | 80 metres maximum | 42 vehicles over 3 minutes | Signing as per page 56-57 END plates if over 50 metres |
| Stop/Go boards | 60 | 100 metres 200 metres 300 metres 400 metres 500 metres | 70 vehicles/ 3 mins 63 vehicles/ 3 mins 53 vehicles/ 3 mins 47 vehicles/ 3 mins 42 vehicles/ 3 mins | Signing as per page 58-59 Consult your supervisor if at or near a railway level crossing See also page 70 |
| Portable traffic lights | 60 | 300 metres maximum | No limit | Advise Highway Authority. Signing as per page 60-61 Consult your supervisor if at or near a railway level crossing See also page 70 |
| Stop-Works Sign | 60 | Not Applicable | Not Applicable | Max period - 2 mins See page 62 |

### Traffic control by 'Give and Take' system

Only use **'Give and Take'** when ALL of the following apply :

- the speed limit is 30 mph or under
- the length of the works from the start of the lead-in taper to the end of the exit taper is 50 metres or less
- drivers approaching from either direction can see 50 metres beyond the end of the works
- two-way traffic flow is less than 20 vehicles counted over 3 minutes (400 veh/hr)
- less than 20 heavy goods vehicles pass the works per hour

The signing you will need is shown on page 55.

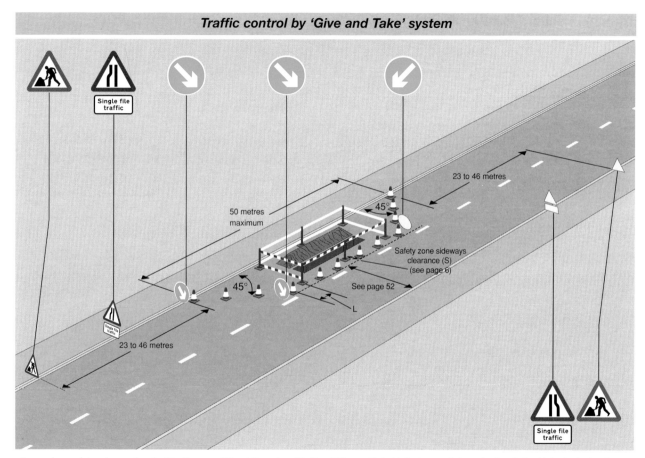

For numbers and size of cones, length of lead in taper (T) and dimensions 'D', 'L' and 'S' see table inside back cover. Information boards should also be displayed (although omitted here for clarity). See page 20.

## Traffic control by Priority signs

Only use **Priority** signs when ALL of the following apply:

- two-way traffic flow is less than 42 vehicles counted over 3 minutes (850 veh/hr)
- the length of the works from the start of the lead-in taper to the end of the exit taper is 80 metres or less
- drivers approaching from either direction can see past the works from a point 60 metres before the coned area to a point 60 metres beyond the coned area in a 30 mph speed limit.
On roads with a higher speed limit the clear visibility distance is:
40 mph  –  70 metres
50 mph  –  80 metres
60 mph  –  100 metres

Face the priority signs in opposite directions. The signs must be placed together with the relevant plate. The sign and its plate 'Give way to oncoming vehicles' must be positioned on the same side of the road as the works.

The signing you need is shown on page 57.

There is no need to use 'End' plates when the single file lane is less than 50 metres long.

## Traffic control by priority signs

For numbers and size of cones, length of lead in taper (T) and dimensions 'D', 'L' and 'S' see table inside back cover. Information boards should also be displayed (although omitted here for clarity). See page 20.

57

## Traffic control by Stop/Go boards

You can control traffic manually by using **Stop/Go** boards when the two-way traffic and the length of the works do not exceed the following :

| Works length (metres) | Maximum two-way traffic flow | |
|---|---|---|
| | Vehicles per 3 minutes | Vehicles per hour |
| 100 | 70 | 1400 |
| 200* | 63 | 1250 |
| 300 | 53 | 1050 |
| 400 | 47 | 950 |
| 500 | 42 | 850 |

\* limit of remotely operated boards

You will only need one board positioned at one end or in the middle if the shuttle lane is 20 metres long or less and the board is clearly visible from both directions, otherwise use a board at each end. The operator showing 'GO' to traffic should be the one to change the direction of traffic flow by reversing the board to show 'STOP'. Adequate time must be allowed for vehicles to clear before the other board is reversed to show 'GO'. Two-way radio control between operators may be needed where the operators are not clearly visible to each other and there is no intermediate operator present.

If traffic flow is not more than 850 vph then remotely operated Stop/Go boards may be used, but only during daylight hours, provided the operator has an unobstructed view of both approaches and is less than 100 metres from both boards.

When the 'Stop/Go' boards are to be used at or near a railway level crossing consult your supervisor (see page 70).

Consult your supervisor if there is a road junction in the shuttle lane, as the use of Stop/Go boards may not be appropriate. The signing you will need is shown on page 59.

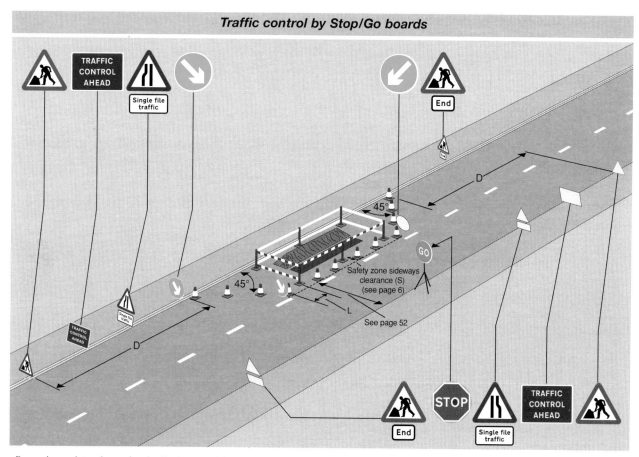

For numbers and size of cones, length of lead in taper (T) and dimensions 'D', 'L' and 'S' see table inside back cover. Information boards should also be displayed (although omitted here for clarity). See page 20.

## Traffic control by portable traffic signals

You can use **portable traffic signals** at most sites up to 300 metres long. Type Approved equipment must be used and should always be vehicle actuated except where otherwise instructed in writing by the highway authority. Your supervisor must tell the highway authority when portable traffic signals are to be used. Approval in writing will be needed when the shuttle section includes a road junction.

Make sure that the exit from the shuttle section does not become blocked by tailbacks. Also ensure that tailbacks from the signals will not block back to a railway level crossing (see page 70). Under no circumstances should portable traffic signals be used at works which straddle a railway level crossing, nor to control road traffic within 50 metres of a level crossing equipped with twin red light traffic signals.

Consider using two traffic signals on each approach, especially where traffic approaches at high speed. In some circumstances this may not be physically possible or necessary. In such cases the normally preferred position for a single traffic signal is at the nearside of the carriageway.

However, there may be good reasons for placing it in the carriageway adjacent to the works. Ensure that there is clear visibility, of at least one signal head, for approaching vehicles. Where power cables cross the carriageway, 'Ramp' signs should be used where the cable protector exceeds 15mm in height.
See the 'The use of vehicle actuated portable traffic signals' (the 'Pink Book') for setting up and adjusting the timings of portable traffic signals. Remember to allow for cyclists and horse riders who travel slower than motor vehicles.

Where a road junction enters in the shuttle section and is not under signal control, a 'Traffic under signal control' sign is required in the joining road, and 'Joining traffic NOT signal controlled' in the main road on the approaches to the junction.

You must have 'Stop/Go' boards available in case the portable traffic signals break down.

The signing you will need is shown on page 61.

## Traffic control by portable traffic signals

For numbers and size of cones, length of lead in taper (T) and dimensions 'D', 'L' and 'S' see table inside back cover. Information boards should also be displayed (although omitted here for clarity). See page 20.

### Short duration Traffic Control by Stop-Works sign

The 'STOP-WORKS' sign may only be used to stop traffic for a short period during works on or near a road. It must not be used as a substitute for other forms of control and should only be used at sites where the risk is assessed as being low. It is used in a similar manner to the School Crossing Patrol sign.

The sign must be double sided and mounted on a black/yellow banded pole, and held by the operator who must be wearing high visibility clothing. The sign must be illuminated when used at night. Two STOP-WORKS signs may be required in circumstances such as manoeuvring plant or works vehicles.

Only use the STOP-WORKS sign when ALL the following apply:
- on single carriageway roads
- when the stoppage is to be for a maximum period of 2 minutes
- the minimum clear visibility for drivers to the sign is
  - 60 metres for speed limits of 40 mph or under
  - 75 metres for speed limits of 50 and above

Unless the site is already signed and guarded, additional signs 'Traffic Control Ahead' must be positioned on both approaches when ANY of the following conditions apply:
- the two-way traffic is greater than 20 vehicles counted over 3 minutes (400 veh/hr)
- bends in the road or other obstructions affect visibility
- the speed limit is 50 and above.

See table inside back cover for the siting distances for these signs.

## SPEED CONTROL

The use of **speed control** as a traffic management option will need to be considered as part of the works planning, as it may not be reasonably practicable to provide full safety zone clearances to suit existing speed limits of 50 or 60 mph. In such cases it may be possible to provide the safety zone clearances applicable to a reduced speed limit of 30 or 40 mph.

To proceed with this option, at the earliest opportunity your supervisor must consult the highway authority who will if necessary impose a temporary mandatory speed limit. This is essential if delays are to be avoided and the order is to be effective. Advice of its existence is given by displaying the appropriate speed limit and 'reduced speed limit ahead' signs as part of the normal signing.

For a standard two-way road with shuttle working as shown on page 64, the length of road covered by the temporary mandatory speed limit should include at least one chicane, and only positive types of traffic control shall be allowed, i.e. Stop/Go boards or portable traffic signals.

In exceptional circumstances where the road width prevents the provision of the appropriate sideways clearance, and diversion of traffic would be impracticable, traffic speeds must be reduced to less than 10 mph and a safe method of working imposed. This must be agreed with the highway authority. At least two chicanes are required, of the minimum size to allow a large vehicle to pass through slowly. Traffic must first be brought to a halt by positive traffic control and then released in small batches by careful use of Stop/Go boards or manually controlled portable traffic signals. This may have the effect of reducing speeds to 10 mph over short lengths of shuttle working. See page 65. For longer lengths or problem sites a convoy system will be necessary, i.e. where a suitable works vehicle leads traffic through at 10 mph. This is the only way of ensuring traffic complies with a 10 mph traffic order. Display 'Convoy system in operation' signs at each end. If convoy working is used your supervisor will need to consult the highway authority and follow the advice set out in the Highways Agency's Advice Note TA63/97.

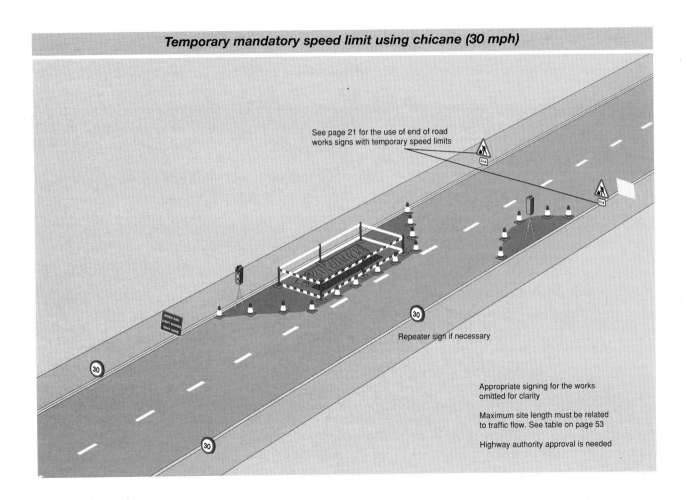

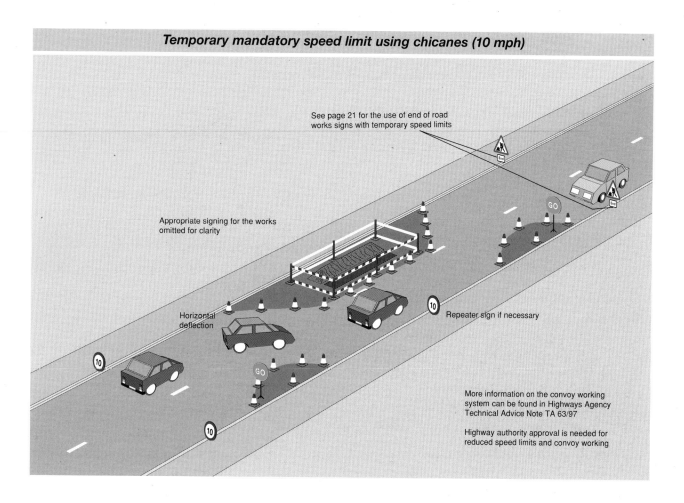

## MOBILE WORKS AND MINOR WORKS CARRIED OUT FROM A VEHICLE

These include continuous **mobile operations**, as well as those which involve movement with periodic stops and short duration static works. They also include **minor works** which do not include excavations, involving the use of a single vehicle or a small number of vehicles.

Works in this category may omit the use of cones and a traffic barrier (lane closed sign) provided that safe working methods are used.

Single vehicle works must not be carried out on dual carriageways to which the national speed limit applies, unless they can be done at prevailing traffic speeds.

### *Using a single mobile vehicle or minor works with one or more vehicles*
Carry out the work when there is good visibility and **during periods of low risk**. Consult your supervisor if work is to take place in the centre of the carriageway with traffic passing on both sides.

Basic requirements :
- the vehicle must be conspicuously coloured
- the vehicle must have one or more roof mounted beacons operating
- a 'Keep Right/Left' sign must be displayed for drivers approaching on the same side of the carriageway, showing which side to pass. Vehicle mounted 'Keep Right/Left' signs must be covered when the vehicle is travelling to and from the site. Do NOT simply turn the sign to point up or down.

### Additional static signs
will be required when ANY of the following conditions apply :
- the works vehicle cannot be seen clearly because of hills, bends in the road, etc.
- stationary traffic may tail back
- there is not enough space for two-way traffic to pass the works vehicle
- the vehicle is slow moving or is required to make periodic stops

In these cases place 'Road Works Ahead' signs with appropriate plates for drivers approaching in each direction.

Where appropriate, you must also display 'Road Narrows' signs with 'Single file traffic' plates.

A 'Road Works Ahead' sign should be displayed to drivers approaching on a side road if work is taking place near a junction.

Mobile works should not be carried out more than 1 mile from these signs.

If any of these basic requirements are not met, you must use full standard signing and guarding.

### *Mobile lane closure on high speed dual carriageways*
When the works cannot be carried out at normal speeds on high speed dual carriageways a mobile lane closure technique may be appropriate. Consult your supervisor if you need to use this technique.

## WORK NEAR TRAMWAYS

Special safety precautions must be taken when works are to be carried out near a tramway. A summary of the main safety points is given below. Detailed advice must be obtained by your supervisor from the relevant track or transport authority prior to starting work and given to those carrying out the works.

### Risk of collision with the tramcar

Unlike other traffic a tramcar cannot swerve to avoid a person or obstruction. Tramcars are wider than the tracks on which they run. The path of a tramcar which must be left unobstructed is known as the 'swept path'. In some cases this is indicated by a line of yellow discs, a painted line or a raised kerb.

It is essential that signing and guarding equipment, operatives, vehicles and pedestrians are kept out of the swept path. Where the works cause the footway to be diverted into the carriageway, the barrier between the pedestrians and the tramway must be kept at least 0.5 metres away from the edge of the swept path.

Where the safety zone sideways clearance would intrude on the swept path, your supervisor should consult the transport authority. The safety zone may be reduced to 300mm and the transport authority may impose a speed restriction on tramcars, and/or provide a lookout.

### Risk of electrocution

Tramway electrical cables consist of overhead lines and underground cables that may be placed outside the swept path.

Your supervisor should liaise with the track or transport authority before working close to overhead lines.

No equipment, plant, vehicles, etc. should be brought within 2 metres of the overhead lines.

Underground cables should be dealt with using standard safe digging practices.

### Tramway crossings

Where a tramway runs on a reserved track but crosses the road at certain places, such crossings should be treated as railway level crossings. See page 70.

## Works adjacent to tramways

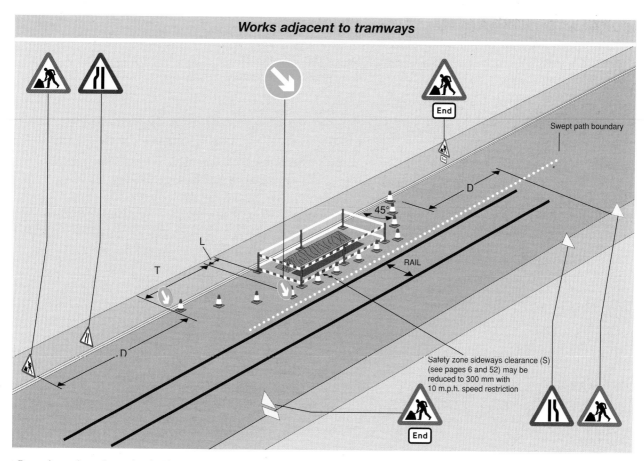

For numbers and size of cones, length of lead in taper (T) and dimensions 'D', 'L' and 'S' see table inside back cover. Information boards should also be displayed (although omitted here for clarity). See page 20.

## WORKS AT OR NEAR RAILWAY LEVEL CROSSINGS

Extreme care must be taken to avoid stationary traffic tailing back across a railway level crossing when street works or road works are being carried out at or near the crossing. Particular attention must be paid to situations where works, even though they may be a considerable distance from the crossing, may cause traffic to tail back over the crossing as a result of long traffic delays. Road traffic must NEVER be stopped on a level crossing. Your supervisor must contact the railway owner when works are to take place at or near a level crossing, or where traffic queues could affect a level crossing.

Detailed advice on carrying out works on or near railway level crossings is given in Appendix F of the *Code of Practice for the Co-ordination of Street Works and Works for Road Purposes and Related Matters*.[1]

This must be given to, and understood by, everyone proposing to carry out works at or near to a railway level crossing.

[1] ISBN 0 11 552310 3  £14  At the time of going to press, this code is applicable to England only.

**Reminder**  (This page is not part of the Code of Practice)

**Before you start – General**
Is high visibility clothing being worn by everyone on site?
Are all signs, barriers, cones and lighting correctly placed?
Are signs obscured by bends, hills or dips in the road?
Are advance signs needed?
Will the site be safe at night or in wind, fog, snow or rain?
Are parked vehicles, trees, street furniture obscuring signs?
Is there enough road width remaining for two-way traffic?
Is traffic control with shuttle lane working required?
Are there any site specific risks requiring special guarding?
Has allowance been made for delivery and removal of materials?
Is the contact number displayed on the information board?

**Before you start – Pedestrians**
Are pedestrians given protected routes which are wide enough?
Are pedestrian routes clearly indicated?
If the footway is closed, is there an alternative route? If so, is it clearly marked?
Are there any special hazards for disabled pedestrians? If so, how can they be made safe?
If a temporary footway in the road is to be used, are ramps to the kerb provided where necessary?

**Before you start – Traffic**
Is type of traffic control right for work, traffic and speed?
Have any misleading permanent signs and road markings been covered?
Is there safe access to adjacent premises?
Have you a copy of portable traffic signals site approval?
Have you considered the needs of cyclists and horse riders?

**When work is in progress**
Does signing and guarding meet changing conditions?
Are signs, cones and lamps being kept clean?
Can traffic control arrangements be improved to reduce traffic delays as conditions change?
Are the carriageway and footway being kept clear of mud and surplus equipment?
Are materials that are left on verges or lay-bys being properly guarded and lit?

**When work is suspended**
Will checks be made on signing, lighting and guarding?
Has the arrangement been changed to reflect conditions?

**When work is finished**
Have all signs, cones, barriers, and lamps been removed?
Have any covered permanent signs been restored?
Have the authorities been told the work is completed?

# INDEX

advance signs 15, 22, 23, 30, 47, 49
amber beacons 4, 22, 23, 34, 47, 66
barriers, pedestrian 18, 20, 29
barriers, traffic 18, 19, 34, 66
basic signs 12
basic site layout 8, 9, 10, 11
buses 52
changing traffic conditions 4
Chapter 8 Traffic Signs Manual iv, v, 1, 38
clearing up 4
cone size ibc
cone taper 5, 7, 34, 52, ibc
cones 17
congested road 7
conspicuous vehicle 19, 34, 66
convoy system 63
cycles 42, 51, 60
dual carriageway 38
emergencies 4
end sign 21
excavations 19, 32, 33
exit taper 6
footway boards 32
footway ramp 32
footway temporary 19, 28, 31
give and take 54
guide island 38
high visibility clothing 2, 15, 23
information board 20

junctions 3, 44, 47, 58, 60
junctions with traffic lights 47
keep right/left sign 16
lamps 17, 19, 22
lead-in taper 5, 7, 34, ibc
lighting 26
longways clearance 5, 35, ibc
maintenance of site 4
minor works 20, 66
mobile works 20, 66
parking 22, 34, 35
pedestrian barriers 18, 20, 29
pedestrian crossings 29, 42
pedestrian safe route 19, 28, 31
pedestrian temporary route 19, 28, 31
pedestrianised areas 29
plates used with signs 14, 16, 56
portable traffic signals 60
priority signs 56
protective clothing 2
railway level crossing 58, 60, 70
ramps 32, 33, 60
reflectorisation 3, 18, 26, 27
risk assessment iv, 2
road narrows sign 16
road plates 33
road widths 6, 35, 52
road works ahead sign 16
roundabouts 49

*ibc = inside back cover*

safety zone 5, 6, 29, 63, ibc
scaffolding 29
securing signs 2
setting out signs 22–25
shuttle lane working 52, 60, 63
sideways clearance 6, 7, 29, ibc
sign lighting 26
signs-duplicates 4
single carriageway 34
speed control 63
speed limit, temporary 7, 63
speeding 7
stop-works sign 26, 27, 62
stop/go boards 26, 58, 60
street lighting 17, 26
tapping rail 18, 20
traffic barriers 18, 19, 34, 36
traffic control 52
traffic signals, permanent 47
traffic signals, portable 60
traffic signs, permanent 2
tramways 68, 69
two-way roads 3, 34
two-way working 52
unobstructed width 52
weather, bad 2, 3, 17
work site access 22
working space 5, 29
works area 5
works vehicle 22, 34, 35